Biting Flies

Biting Flies

Patrick Merrick

THE CHILD'S WORLD®, INC.

Library of Congress Cataloging-in-Publication Data
Merrick, Patrick.
Biting flies / by Patrick Merrick.
p.   cm.
Includes index.
Summary: Describes the physical characteristics, behavior,
and life cycle of different kinds of flies that bite.
ISBN 1-56766-631-0 (lib. bdg. : alk. paper)
1. Flies—Juvenile literature.
[1. Flies.]   I. Title.
QL533.2.M47   1999
575.77—dc21   98-45791
CIP
AC

Photo Credits

© Alvin E, Staffan, The National Audubon Society Collection/Photo Researchers: 16
ANIMALS ANIMALS © George Bryce: 6, 26
© Bill Beatty/Wild & Natural: 9
© Dwight R. Kuhn: cover, 2, 19, 20, 23, 24
© 1999 E.R. Degginger/Dembinsky Photo Assoc. Inc.: 29
© Joe McDonald/Visuals Unlimited: 30
© Rick Poley/Visuals Unlimited: 15
© Robert and Linda Mitchell: 10
© William E. Ferguson: 13

On the cover...

Front cover: From close up, you can see the detail of this *horse fly's* eyes.
Page 2: This *deer fly* is resting on a leaf.

Table of Contents

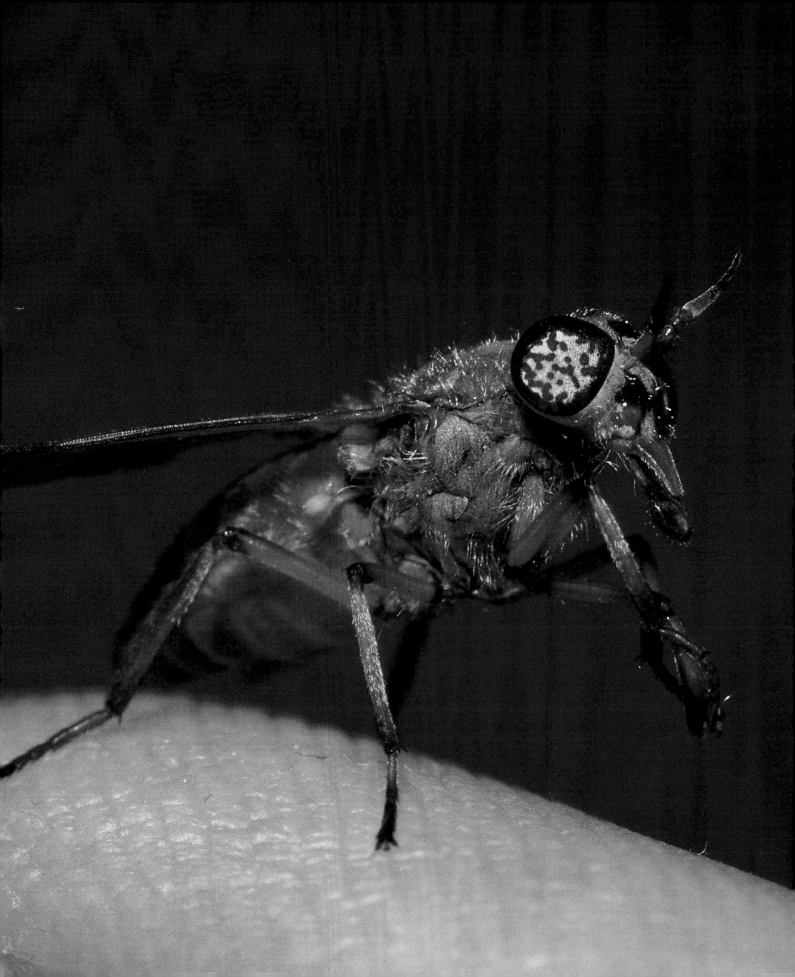

Summertime is a wonderful time of the year. It brings warm sunshine, green grass, and gentle breezes. Sometimes, summer brings things that bother us, too. Bees, ants, and other bugs sometimes bother people who are enjoying the outdoors. Once in a while, bugs such as flies land on people and animals. Some of these flies bite their victims and drink their blood. What are these summertime pests? They are biting flies!

⇐ This *horse fly* is on the photographer's finger.

Are There Different Types of Flies?

Biting flies are not really scary creatures. In fact, flies are some of the most common kinds of **insects.** An insect is an animal that has three body parts: a head, a **thorax,** and an **abdomen.** The thorax is the insect's chest. The abdomen is its stomach. Insects also have six legs and one or two pairs of wings.

This *deer fly* is getting ready to bite someone's arm. ⇒

There are many different kinds, or **species,** of flies. While some kinds of flies are helpful to people, most are considered harmful. Some types just bother us, while others can actually destroy fruits, meats, and other foods. Some of the most troublesome flies are the biting flies. The *horse fly,* the *deer fly,* and the *sand fly* are examples of biting flies.

What Do Biting Flies Look Like?

There are more than 2,000 different species of horse, deer, and sand flies. Horse flies are the largest—about the size of a bumblebee. They have dark wings and thick, dark bodies. Deer flies look a lot like horse flies, but they are smaller and have lighter wings and bodies.

Sand flies are the smallest of the biting flies. Even when they are adults, sand flies are still no bigger than a grain of sand! Like horse flies, sand flies have dark bodies and dark wings. All three of these flies have huge, brightly colored eyes. In fact, their eyes are so big that they almost cover the fly's whole head!

It's easy to see how tiny this *sand fly* is as it walks on a person's arm. ⇒

What Do Biting Flies Eat?

Biting flies are fast, strong fliers. They also have very sharp mouth parts. Male biting flies feed mostly on plant juices. Female biting flies use their sharp mouths to feed on something very different—blood!

When they are hungry, female biting flies fly up to a person or other animal. They look all over for a place to feed. Most of the time, they choose the person's arms, head, or neck. Once a biting fly has picked a spot, it bites a small hole in the skin. As blood comes out of the hole, the fly drinks it up. When it is full, the fly simply stops drinking and flies away.

This female deer fly is almost ready to bite this person's hand. ⇒

Biting flies live all over the world. They are very common in the United States and in Canada. Most of the time, they are found in forests or in wet areas near ponds, lakes, and rivers.

Biting flies come out to feed during the day. Their favorite days are when the weather is very hot and sticky. Biting flies do not like to fly in a lot of wind, so they feed on days when the air is very still.

⇐ This horse fly is resting on a leaf in Ohio.

How Are Biting Flies Born?

When male and female biting flies mate, the female must feed on blood so she can make her eggs. When she is ready to lay her eggs, she finds some water or wet ground. Then she lays her eggs over that spot.

In about a week, the tiny eggs hatch and the baby biting flies fall into the water. A baby biting fly is called a **larva.** It does not look like a fly at all. In fact, it looks like a very small worm!

In this picture, *black fly* eggs and larvae are underwater in a stream. ⇒

When the larva hatches, the only thing it wants to do is eat. Biting fly larvae will eat anything. They will try to make a meal of plants, animals, other flies, and even people!

As the larva eats, it grows bigger. The larva doesn't grow in the same way people do. When it gets too big for its skin, it **molts,** or sheds the old skin. When a larva molts, there is always a new and bigger skin underneath.

Finally, after the larva has grown as big as it can, it is ready to **pupate,** or change. When a larva pupates, its body rebuilds itself into a new shape. The old wormlike shape slowly changes into a fly. The thorax and abdomen appear, and the larva grows wings. Then the larva molts one last time. After this last molt, the larva is an adult fly.

This adult black fly is resting on the water after pupating. ⇒

Do Biting Flies Have Enemies?

Lots of animals like to eat biting flies. As small larvae, flies are eaten by fish and other water creatures. Frogs, birds, and even other insects like to have a meal of adult biting flies. People are also a big enemy to biting flies. Since the bites itch and bother us, we often try to kill biting flies with sprays or hit them with swatters.

⇐ This big *grass fly* is feeding on a black fly that it caught.

Are Biting Flies Dangerous?

Since biting flies make a hole in your skin to drink blood, their bites sometimes hurt. When the fly is finished drinking, its bite often itches and swells up, too. Sometimes, biting flies can make people very sick. That's because some flies carry **diseases,** or illnesses. When these flies bite people or animals, their victims can become very sick.

How Can We Stay Safe from Biting Flies?

Since biting flies live all over the world, it is difficult to avoid them. Even so, you can protect yourself by following some simple rules. First, when you go to areas where biting flies might live, wear long pants and a shirt with long sleeves. This way, the flies can't get to your skin to bite you.

You can also use bug spray. Spraying it on your clothes will keep flies away from you—sometimes for several hours! Always make sure you have an adult with you when you use bug spray.

It's easy to see the beautiful colors of this deer fly's eyes. ⇒

Most people think biting flies are mean creatures that bite people to hurt them. This is not true. Biting flies bite people for the same reason you eat a hamburger—to get enough food to live! If we remember this and protect ourselves from their bites, we can enjoy summer without worrying about biting flies.

Glossary

abdomen (AB–duh–men)
The abdomen is the stomach part of an insect.

diseases (dih–ZEE–zez)
Diseases are illnesses. Some biting flies carry diseases and pass them on when they bite someone.

insects (IN–sekts)
An insect is an animal that has a body with three parts. Insects also have six legs.

larva (LAR–vah)
A larva is a young insect. Biting fly larvae look like little worms.

molts (MOLTS)
When an animal molts, it sheds its old skin. Underneath is a newer, bigger skin that lets the animal grow larger.

pupate (PYOO–payt)
When an animal pupates, its body changes shape. When a biting fly pupates, it changes from a wormlike larva to an adult fly.

species (SPEE–sheez)
A species is a different kind of an animal. There are many different species of biting flies.

thorax (THOR–ax)
The thorax is the chest of an insect.

Web Sites

Learn more about biting flies:

http://www.ifas.ufl.edu/~insect/livestock/deer_fly.htm

http://content.ag.ohio-state.edu/ohioline/hyg-fact/2000/2115.html

http://www.oznet.ksu.edu/Dp_entm/extension/InsectID/Mock/tabanid.htm